U.S. Department
of Transportation
National Highway
Traffic Safety
Administration

DOT HS 809 760   September 2004

Technical Report

An Assessment of the Crash-Reducing Effectiveness of Passenger Vehicle Daytime Running Lamps (DRLs)

Published By:

NCSA

National Center for Statistics and Analysis
Advanced Research and Analysis

1. Report No. DOT HS 809 760	2. Government Accession No.	3. Recipient's Catalog No.
4. Title and Subtitle An Assessment of the Crash-Reducing Effectiveness of Passenger Vehicle Daytime Running Lamps (DRLs)		5. Report Date September 2004
		6. Performing Organization Code NPO-101
7. Author(s) Joseph M. Tessmer		8. Performing Organization Report No.
9. Performing Organization Name and Address Mathematical Analysis Division, National Center for Statistics and Analysis National Highway Traffic Safety Administration U.S. Department of Transportation NPO-121, 400 Seventh Street, S.W. Washington, D.C. 20590		10. Work Unit No. (TRAIS)
		11. Contract or Grant No.
12. Sponsoring Agency Name and Address Mathematical Analysis Division, National Center for Statistics and Analysis National Highway Traffic Safety Administration U.S. Department of Transportation NPO-121, 400 Seventh Street, S.W. Washington, D.C. 20590		13. Type of Report and Period Covered NHTSA Technical Report
		14. Sponsoring Agency Code

15. Supplementary Notes

Abstract

This study estimates the effectiveness of passenger vehicle daytime running lights in reducing two-vehicle opposite direction crashes, pedestrian/bicycle crashes, and motorcycle crashes. The authors chose the generalized simple odds, a conventional statistical technique, to analyze the data.

Results based on simple odds indicate that from 1995 to 2001:

- DRLs reduced opposite direction daytime fatal crashes by 5 percent.

- DRLs reduced opposite direction/angle daytime non-fatal crashes by 5 percent.

- DRLs reduced non-motorists, pedestrians and cyclists, daytime fatalities in single-vehicle crashes by 12 percent.

- DRLs reduced daytime opposite direction fatal crashes of a passenger vehicle with a motorcycle by 23 percent.

Reviewers of this paper required the inclusion of results using the odds ratio technique. The estimated the effect of DRLs are −6.3 percent, −7.9 percent, 3.8 percent, and 26 percent, respectively. None of these results were statistically significant.

17. Key Words daytime running lamps, daytime running lights, DRL, DRLs, logistic regression, simple odds, fatal crashes, FARS, NASS, GES, motorcycle, pedestrians, cyclists	18. Distribution Statement Document is available to the public through the National Technical Information Service, Springfield, VA 22161 http//:www.ntis.gov
19. Security Classif. (of this report) Unclassified	20. Security Classif. (of this page) Unclassified	21. No. of Pages 36	22. Price

Form DOT F 1700.7 (8-72)   Reproduction of complete page authorized

TABLE OF CONTENTS

NCSA   National Center for Statistics and Analysis ♦ 400 Seventh St., S.W., Washington, D.C. 20590

List of Tables

Executive Summary

This study estimates the effectiveness of passenger vehicle daytime running lights in reducing two-vehicle opposite direction crashes, pedestrian/bicycle crashes, and motorcycle crashes. The authors chose the generalized simple odds, a conventional statistical technique, to analyze the data. The generalized odds ratio attempts to adjust for a variety of exogenous factors other than the presence or absence of DRLs not specifically controlled for within the model.

Significant results of this study show that from 1995 to 2001:

Simple Odds Results:
- DRLs reduced opposite direction daytime fatal crashes by 5 percent.

- DRLs reduced opposite direction/angle daytime non-fatal crashes by 5 percent.

- DRLs reduced non-motorists, pedestrians and cyclists, daytime fatalities in single-vehicle crashes by 12 percent.

- DRLs reduced daytime opposite direction fatal crashes of a passenger vehicle with a motorcycle by 23 percent.

The reviewers of this paper required the inclusion of an analysis based on odds ratio, which can be found in Appendix B. Like the simple odds, the odds ratio attempts to control for a variety of factors other than the presence or absence of DRLs. The estimated effectiveness of DRLs based on this technique is extremely sensitive to small changes encountered in real world crash data. As a result, reductions in target crashes during the daytime using the odds ratio technique may not be detected over the inherent background noise of the data system. **None** of the results based on the odds ratio are statistically significant.

Odds Ratio Results:
- DRLs reduced opposite direction daytime fatal crashes by –6.3 percent that is DRLs increase opposite direction daytime fatal crashes by 6.3 percent.

- DRLs reduced opposite direction/angle daytime non-fatal crashes by –7.9 percent that is DRLs increase opposite direction/angle daytime non-fatal crashes by 7.9 percent.

- DRLs reduced non-motorists, pedestrians and cyclists, daytime fatalities in single-vehicle crashes by 3.8 percent.

- DRLs reduced daytime opposite direction fatal crashes of a passenger vehicle with a motorcycle by 26 percent.

METHODOLOGY:

A case-control method was chosen as the approach for this study, since only specific make-models for each year were equipped with DRLs. The number of crashes for a set of passenger vehicles equipped with DRLs is compared to passenger vehicles manufactured in the same years without DRLs. The groups of vehicles are analyzed by time of day and crash type.

The generalized simple odds method was used to analyze the data. This technique implicitly attempts to control for factors, other than the presence or absence of DRLs, that could be associated with crash occurrences. The effectiveness of DRLs due to differences in passenger vehicle types, namely, passenger cars, SUVs, vans, and light/pickup trucks is addressed explicitly. The simple odds provided useful statistically significant results.

Background

This is the second NHTSA study on the effectiveness of Daytime Running Lamps (DRLs). The preliminary study was published in June 2000 and is the basis of this research.

Many traffic crashes are the result of the failure of a driver to notice another vehicle. Visual contrast is an essential characteristic that enables a driver to detect vehicles. The purpose of daytime running lamps (DRLs) is to increase the drivers' ability to detect DRL-equipped vehicles, particularly in the peripheral visual field, by increasing visual contrast. Seven countries require the use of DRLs during all daytime periods: Canada, Denmark, Finland, Hungary, Iceland, Norway, and Sweden. Results of DRL studies from these countries consistently, however not conclusively, show that DRLs reduce the number of two-vehicle crashes during daylight, dusk, and dawn. This study examines the effectiveness of first-generation DRLs, using U.S. national data for passenger vehicles.

DRLs come in a variety of configurations. DRLs may be upper beam headlamps at reduced intensity, low-beam headlamps at full or reduced power, turn signals or dedicated lamps. In addition the brightness, color and light dispersion are design features of DRLs. Four manufacturers began equipping selected 1995 model year vehicles, for sale within the U.S., with DRLs. General Motors Corporation produces DRL-equipped vehicles with higher intensity DRLs than those used in Scandinavian countries. In the U.S. the availability of DRL-equipped vehicles has increased with each model year since 1995. Since the cost of DRLs is low, small reductions in the number of crashes would likely be considered cost effective. A partial chronological summary of results from several previous studies of the effectiveness of DRLs follows.

Finland's legislation of 1972 required the use of low-beam headlights in rural areas during winter. The rural multiple-vehicle daytime crash rate decreased by 27 percent as a result.[1]
In 1975, Clayton and Mackay[2], at Indiana University, found that drivers failing to process information properly caused almost half of all crashes. The most prevalent information processing errors were faulty visual perception, recognition errors and comprehension errors. In addition, it was shown that traffic crashes were due more to inattention and distraction than to poor vision. The crash reduction potential of DRLs lies in their ability to attract attention, especially in the peripheral visual field, thereby enhancing detect ability.

A study conducted by Transport Canada[3] in 1975-1976 examined the crash experience with part of the Canadian defense vehicle fleet equipped with automatic headlights, a version of DRLs. The results

---

[1]Andersson, K., Kilsson, G., and Salusjärvi, S. The Effect On Traffic Accidents on the Recommended use of Vehicle Running Lights in the Daytime in Finland. Report No 102. Swedish road and Traffic Research Institute (VTI), 1976.

[2]Claton, A.B. and Mackay, G.M. Aetiology of Traffic Accidents. Health Bulletin, 31(4), 277-280, 1972.

[3]Attwood, D.A. The Potential of Daytime Running Lights as a Vehicle Collision Countermeasure. SAE Technical Paper 810190. Society of Automotive Engineers, 1981.

published by Attwood in 1981 showed a 20 percent crash decrease in the specially equipped vehicles compared to the comparison group of unmodified vehicles.

Swedish legislation required the use of DRLs throughout the year starting in October 1977. An 11 percent reduction in daytime crashes was observed. Two-vehicle, head-on crashes were reduced by 10 percent, angle crashes were reduced 9 percent, crashes involving a bicycle or moped were reduced by 21 percent, and crashes involving a pedestrian or a cyclist decreased 17 percent.[4] These results were questioned by Theeuwes and Riemersma in 1995[5], as the proportion of multi-party crashes was not reduced as a proportion of all crashes.

Hills, in 1980[6], and more recently Sekuler and Blake,[7] found that increasing the visual contrast of a vehicle increases the ability of other drivers to detect and monitor the vehicle. Low contrast between a vehicle and its background can be quite common during daylight hours. Contrast is reduced by color, rain, clouds and low levels of light that occur at dawn and dusk.

Stein reported in 1985[8] the results of a study by the Insurance Institute for Highway Safety (IIHS), which equipped over 2,000 passenger cars, light trucks and vans with DRLs. Relevant multi-vehicle crashes were 7 percent lower for the DRL-equipped vehicles than the comparison (unmodified) vehicles.

Norway required the installation of DRLs by vehicle manufacturers in January of 1985 and the use of low beam head lights was required on all vehicles in Norway not equipped with DRLs in April of 1988. Elvik reported[9] that a 15 percent reduction in all summertime multi-vehicle daylight crashes was achieved.

Canada required that all new passenger cars, trucks, multi-purpose vehicles, and buses manufactured for sale in Canada be equipped with DRLs after December 1, 1989. In September 1993 Arora, et al.[10]

---

[4]Andersson, K. Nilsson, G. The Effects on Accidents of Compulsory Use of Running Lights During Daylight in Sweden. Report No. 208A, Swedish Road and Traffic Research Institute (VRI),1981.

[5]Theeuwes, J. and Riemersma, J. Daytime Running Lights as a Vehicle Collision Countermeasure: The Swedish Evidence Reconsidered. Accident. Anal. Prevention. 27:633-642, 1995.

[6]Hills, B.L. Vision, Visibility and Perception in Driving. Perception, 9, 183-216, 1980.

[7]Sekuler, R. and Blake, R. Perception, (Second Edition) Toronto: McGraw-Hill, 1990.

[8]Stein, H. Fleet Experience with Daytime Running Lights in the United States. SAE Technical Paper 851239. Warrendale, PA, Society of Automotive Engineers, 1985.

[9]Elvik, R. The Effects of Accidents of Compulsory Use of Daytime Running Lights for Cars in Norway. Accident Analysis and Prevention, 25(4) 383-398, 1993.

[10]Arora, H. Collard, D. Robbins, G. Welbourne, E.R. White, J.G. Effectiveness of Daytime Running Lights in Canada, Report No. TP1298 (E), Transport Canada 1994.

conducted an extensive analysis on the effectiveness of DRLs for Transport Canada. They estimate that relevant crashes were reduced by 11.3 percent, which was statistically significant at p<0.05.

In October of 1990, Denmark required universal use of DRLs. No overall effect was reported. However, Hansen identified a statistically significant 37 percent decrease in crashes involving a left turn in 1993[11].

Hungary has required the use of DRLs on rural roads since March 1993. Hollo studied the crash experience of DRL-equipped vehicles and presented the findings at a conference in the Czech Republic in 1995[12]. Several changes in traffic regulations and enforcement, which includes the reduction of the speed limit, stricter seat belt laws, increases in police patrols, significantly higher fines and a campaign to increase public awareness of traffic-related issues were considered confounding factors, thereby making it difficult to estimate the effect of DRLs. Nonetheless, Hollo estimates that DRLs reduced the number of rural daytime "frontal and cross traffic" crashes by 7 to 8 percent. Hallo further claims that during "good visibility" crashes are reduced 11 to 14 percent.

IIHS' Highway Loss Data Institute (HLDI) in 1997[13] released findings from a study of the personal injury claims for vehicles that added DRLs as a standard feature in 1995 and 1996, compared to the claim frequencies for the same makes and models prior to adding DRL. The number of relative claims was found to have increased slightly after DRLs were introduced. However, HLDI's study was not able to identify a consistent pattern of increases among vehicles. HLDI's study hypothesized that this finding was not surprising, as "...claims for striking vehicles, single-vehicle crashes, and nighttime crashes could not be identified..." and therefore, could not be excluded from the study. Striking vehicle, single-vehicle, and nighttime crashes would not likely be impacted by the presence of DRLs.

Tofflemire and Whitehead[14] re-analyzed the Canadian DRL law in 1997 using a "quasi-experimental comparative posttest design" and found that opposite direction and angle crashes were reduced by 5.3 percent, which was statistically significant at p<0.05. The study concluded that the DRL law had a greater effect on opposite direction crashes (15 percent reduction) than angle crashes (2.5 percent reduction).

Each province in Canada was individually analyzed. Only Nova Scotia and New Brunswick

---

[11]Hansen, L.K. Daytime Running Lights in Denmark - Evaluation of the Safety Effect. Translated exact.

[12]Hollo, P. Changes of the DRL-Regulations and their Effects on Traffic Safety in Hungary. Paper presented at the conference: Strategic Highway Safety Program and Traffic Safety, the Czech Republic, September 20-22, 1995. Preprint for sessions on September 21, 1995.

[13]Highway Loss Data Institute Bulletin Volume 15, Number 1, December 1997.

[14]Tofflemire, T. C., Whitehead, P.C. An Evaluation of the Impact of Daytime Running Lights on Traffic Safety in Canada, Journal of Safety Research, Volume 28, Number 4, 1997.

experienced a statistically significant (p<0.05) reduction in crashes.

While the 1993 and 1997 Canadian studies described above are among the few studies reporting statistically significant results, in most other studies the data sets are small, which can result in nonsignificant statistical results, even when an effect might exist.

Hollo[15] reported that DRLs reduced daylight frontal and crossing vehicle crashes by 4.7 percent to 15.2 percent in Hungary, depending on the statistical technique used and assumptions made.

Tessmer[16] estimated that the effectiveness of DRLs in US fatal two-vehicle opposite-direction crashes ranged from –8 percent to 2 percent. For non-fatal crashes the effectiveness ranged from 5 percent to 7 percent. For pedestrians fatalities in single-vehicle crashes, the estimated effectiveness ranged from 28 percent to 29 percent.

Lau[17] estimates that DRLs reduce multiple vehicle crashes by 5 to 13 percent. Lau even estimates that DRLs reduce multiple vehicle nighttime crashes by 5 percent, which suggests that there may a confounding lurking variable within the data.

Farmer and Williams[18] demonstrated that DRLs are associated with a 3.2 percent decline in multiple-vehicle daylight crashes.

Thompson[19] in 2003 presented a paper at the April SAE meeting in Washington, DC. He estimated that DRLs reduced multiple vehicle collisions by 2.3 percent to 12.4 percent, depending on DRL type.

Table 1 summarizes findings from studies of the effectiveness of DRLs in several countries, including the U.S. The individual studies are identified by year, investigator(s), the type of study, i.e., did the study analyze the effects of DRLs on a specific fleet of vehicles, a case controlled study, or the result of a change in the law, applicable country, and the estimated effects of DRLs.

Table 1
Summary of Findings on DRL Effectiveness*

[15]Hollo, P., Changes in the Legislation on the Use of Daytime Running Lights by Motor Vehicles and Their Effect on Road Safety in Hungary, Accid. Anal. And Prev., Volume 30, No.2, pp 183-199, 1998.

[16]Tessmer, J.M., A Preliminary Assessment of the Crash-Reducing Effectiveness of Passenger Car Daytime Running Lamps (DRLs) ; DOT HS 808 645, June 2000.

[17] Lau, E. Daytime Running Light Effectiveness A Preliminary Evaluation, Presented at a Government/Industry Meeting, June 19-21, 2000 Washington, DC.

[18] Farmer, C.M. and Williams, A.F. Effects of daytime running lights on multiple-vehicle daylight crashes in the United States; Accid. Anal. And Prev., Volume 34, pp 197-203, 2002.

[19]Thompson, P.A., Daytime Running Lamps (DRLs) for Pedestrian Protection SAE Paper 2003-0102072, April 2003.

Year	Investigator(s)	Study Type	Country	Estimated Effects
1972	Anderson et al[1]	Law	Finland	27% reduction rural multi-vehicle
1975	Attwood[3]	Fleet	Canada	20% some defense vehicles
1977	Anderson et al[4]	Law	Sweden	9% to 21% crash type dependent
1985	Stein[8]	Fleet	U.S.	7% reduction selected vehicles
1988	Elvik[9]	Law	Norway	15% reduction summer multi-vehicle
1993	Arora et al[10]	Law	Canada	11.3% reduction 2-vehicle opposite-direction
1993	Hansen[11]	Law	Denmark	up to 37% reduction - crash type dependent
1995	Hollo[12]	Law	Hungary	7% to 14% reduction frontal cross traffic
1997	Tofflemire et al[14]	Law	Canada	5.3% reduction opposite direction/angle crashes
1998	Hollo[15]	Law	Hungary	4.7% to 15.2% reduction frontal cross traffic
2000	Tessmer[16]	CC	U.S.	-8% to 29% crash type dependent
2000	Lau[17]	CC	U. S.	5% to 13% reduction multiple vehicle crashes
2002	Farmer et al[18]	CC	U.S.	3.2% decline in mult. vehicle daylight crashes
2003	Thompson[19]	CC	U.S.	2.3% to 12.4% DRL type dependent

* See Bibliography for detailed information on published studies.

Several factors could influence the effectiveness of DRLs, e.g., geography and the climate, the mix of rural and urban crashes, traffic conditions, and manner of collision. The approach of this study attempts to limit the influence of such exogenous variables by using comparison groups where the effects should be similar. This study examines the effectiveness of DRLs in the U.S. for vehicles of model years 1995 and later. Two sources of data maintained by the National Highway Safety Traffic Administration (NHTSA) are used to study DRL effectiveness: the Fatality Analysis Reporting System (FARS) and the National Automotive Sampling System/ General Estimates System (NASS/GES).

Methodological Changes from Preliminary Assessment

This study is the second study conducted by NHTSA to determine the effectiveness of Daytime Running Lamps (DRLs). The same basic statistical techniques to evaluate DRLs have been used. However, with the collection of additional data and the knowledge gained from NHTSA's first study, A Preliminary Assessment of the Crash-Reducing Effectiveness of Passenger Car Daytime Running Lamps (DRLs), which appeared in 2000, several improvements have been made. A great deal was learned about using

national traffic crash data to analyze DRLs, which guided our efforts in the current study.

In the original study two comparison groups of fatal crashes were used, single vehicle fatal crashes and 2-vehicle same direction fatal crashes. There are many more single vehicle fatal crashes than 2-vehicle same direction fatal crashes. The results of the analysis based on using the 2-vehicle same direction fatal crashes do not produce sufficient power to reject the null hypothesis. Critics of the earlier study pointed out that in same direction crashes, a potential striking vehicle with DRLs could have the DRLs detected in the rear view mirror of the potentially stuck vehicle, which could then take corrective action. They argue that same direction crashes are not independent of DRLs and using them, as a comparison group would skew the results. For these two reasons, analysis using 2-vehicle same direction fatal comparison crashes has been eliminated from this study.

In the original study, both the simple odds, $O = TD/(CD+TN+CN)$[20], and the odds ratio, $? = (TD/CD)/(TN/CN)$[1], were used in the analysis. The standard error of the odds ratio is much larger than the standard error of the simple odds. To be statistically precise, when using the simple odds, the null hypothesis can be marginally rejected, however, the power of the odds ratio is not sufficient to reject the null hypothesis. Therefore the analysis in the main body of this report was based solely on the simple odds. Several reviewers of the report required publication of the non-statistically significant results, based on the odds ratio. The results based on the odds ratio can be found in Appendix B of this report. Generalized forms of the simple odds and the odds ratio were also used in this study; see the appendices. A generalized form of the ratios allows one to adjust for a variety of identifiable factors such as vehicle type.

Target vehicles with DRLs and the comparison vehicles without DRLs have been partitioned in a different way. In the original study two groups of comparison passenger cars were used. The groups of target and comparison vehicles were identified by make and model. The original study's first comparison group consisted of vehicles of the same make and model prior to the adoption of DRLs. Vehicles in this comparison group were from 1 to 6 years older than the target vehicles equipped with DRLs. To eliminate the potential bias due to age in the original study a second group of comparison vehicles was selected, namely vehicles manufactured by the Ford Motor Company at the same time that the target vehicles were manufactured.

In the current study, the vehicles under analysis have been expanded from passenger cars to passenger vehicles. Both the target and comparison vehicles have been identified by analysis of the vehicle identification number (VIN). Target and comparison vehicles were all manufactured during the same time period. All passenger vehicles that could be classified as having DRLs as standard equipment were classified as target vehicles. All passenger vehicles that did not have DRLs as standard equipment nor as a standard option were included as comparison vehicles.

The effectiveness of DRLs in preventing fatal two-vehicle daytime opposite direction crashes of passenger vehicles with motorcycles was examined.

---

[20] See Appendix A, Page 19 for additional details on the simple odds and odds ratios.

Four states, Florida, Maryland, Missouri and Pennsylvania were used to examine the effectiveness of DRLs for non-fatal crashes in the original study. However, one cannot extrapolate the effectiveness of DRLs to the nation. To obtain a national estimate, data from the General Estimates System (GES) was used. Since GES is a survey and not a census of crashes, software for the statistical analysis of correlated data, SUDAAN, was used to obtain credible estimates of statistical significance.

Finally a meta-analysis was used in the original study to attempt to provide an overall estimate of DRL effectiveness. This has been eliminated from the current analysis since the survey data was used to estimate the effects of DRLs for non-fatal crashes. The mixture of survey data and census data in a meta-analysis does not provide reliable results.

Data and Methodology

Previous studies of DRL effectiveness often have used a before vs. after approach. This approach is appropriate, for example, when a law goes into effect at a given point in time and one wishes to determine the effect of that law on traffic crashes. A case-control method was chosen as the approach for this study, since only specific make-models for each year were equipped with DRLs. A case-control method attempts to control for factors, other than the presence or absence of DRLs that could be associated with crash occurrence. In this study, the number of crashes for a fleet of vehicles equipped with DRLs is compared to a fleet of vehicles without DRLs produced in the same years. Both groups of vehicles are analyzed by time of day and crash type. Analysis of the Vehicle Identification Number (VIN) was used to partition passenger cars, vans, pickups/light trucks, and sport utility vehicles (SUVs) into a fleet of vehicles that did and did not have DRLs. Passenger vehicles that permitted DRLs as a standard option were removed from the analysis, since one could not analyze the VIN to determine if the specific vehicle was or was not equipped with DRLs.

Data from FARS[21] for calendar years 1995 - 2001 were used to examine DRL effectiveness for fatal two-vehicle opposite-direction crashes and for single-vehicle pedestrian/cyclist crashes. NASS/GES[22] data for calendar years 1995 – 2001 were used to examine DRL effectiveness for non-fatal two-vehicle opposite-direction crashes.

The analysis focused on the possible effect of DRLs in reducing crashes during daylight or twilight hours, as opposed to nighttime hours, when traditional lighting would be in use by all drivers. Therefore, the

---

[21] Fatal crash data are from NHTSA's *Fatality Analysis Reporting System (FARS)*. FARS contains data on a census of fatal traffic crashes within the United States and Puerto Rico. A crash must involve a motor vehicle traveling on a public roadway and must result in the death of an occupant of a vehicle or a non-motorist within 30 days of the crash to be included in FARS.

[22] Non-fatal crash data are from NHTSA's *National Automotive Sampling System/General Estimates System (NASS/GES)*. NASS/GES contains data from a survey of approximately 55,000 weighted traffic crashes across the United States. Both injury crashes and property damage only crashes are included.

target time period is daytime, including dawn and dusk, and the comparison time period is night23.

Target crashes and comparison crashes are defined by the crash configuration. Ideally, the only difference between daytime target crashes and daytime comparison crashes is that the set of daytime target crashes consists of crashes that could be affected by DRLs, while the set of daytime comparison crashes consists of crashes that would not be affected by DRLs. A target crash is a crash where the DRLs can be seen by the driver of the other crash involved vehicle. A comparison crash is a crash involving a single vehicle, where the visibility of DRLs is not relevant.

Neither the FARS nor the NASS/GES databases have a variable that partitions the data exactly into target and comparison crashes. Both data sets have variables, which permit one to approximate the desired partition. Therefore, it is possible that the partition of target crashes and comparison crashes may not be perfect. For example, the geometry of an angle crash might prevent a driver from seeing the DRLs of the other vehicle. If angle crashes that cannot be affected by DRLs are included in the set of target crashes, the estimated effect of DRLs, using FARS may be underestimated. Since the effectiveness is expected to be small, fatal target crashes have been limited to head-on crashes and sideswipe opposite direction crashes. Although the glare from DRLs may contribute to a single vehicle crash, this is unlikely. However, the data do not have the fidelity to identify such crashes. At night, one assumes neither the target crashes nor the comparison crashes should be affected by DRLs. This assumption, like all assumptions can be challenged. For example, if a driver of a DRL-equipped vehicle does not turn on his head/tail lights at night a crash may result. Again this unlikely set of events is within the realm of possibility; however, the available data do not permit one to identify or analyze such crashes. Two-vehicle target crashes were further distinguished, for the purposes of this study, by focusing on those involving crashes in which the two vehicles were traveling in opposite-directions.

The FARS and NASS/GES target crashes include head-on and sideswipe opposite direction crashes[24]. The set of single-vehicle crashes is used as a set of comparison crashes. The comparison groups of crashes, ideally, would represent those crashes, which would not be affected by the presence or absence of DRLs. In the case of nighttime crashes, it has been pointed out that the use of DRLs may cause headlamps to burn out more frequently, contributing to an increase in nighttime crashes. However, only early Volkswagen and Volvo vehicles use full intensity lower beam headlamps for DRLs. In addition, all vehicles equipped with DRLs are relatively new, model year 1995 and later, so the potential problem of burned out headlamps should be minimal. Hauer (1995) pointed out that single-vehicle crashes might also be affected by DRLs. Namely, two-vehicles on a collision course may detect each other earlier due to DRLs. In such a situation, a multi-vehicle crash may be avoided and a single-

---

[23] An alternative partition of the light condition would be to exclude all dawn and dusk crashes from the analysis. A preliminary analysis to calculate the point estimate of DRL effectiveness during dawn and dusk was made. The result showed a larger value of DRL effectiveness during dawn and dusk than during the day. However, due to the limited number of dawn and dusk crashes, the result was not statistically significant.

[24] Sideswipe opposite direction crashes are two-vehicle crashes with the vehicles moving in opposite directions. The initial engagement does not overlap the corner of either vehicle by more than four inches, so that there is no significant involvement of the front or rear surface areas. In addition, there is no pocketing of the impact in the suspension areas. The impact swipes along the surface of the vehicles parallel to the direction of travel. There is low retardation of the force along the surface of the vehicles.

vehicle crash may result. Thus, the two comparison groups, nighttime crashes and single-vehicle crashes may not be statistically independent of DRLs, a required theoretical assumption for the analysis performed here. However, from a practical point of view, these two groups are as statistically independent from the target as is reasonably possible. That is, in general, a two-vehicle opposite-direction crash does not cause, nor does it prevent, a single vehicle crash. Likewise, a single-vehicle crash does not cause, nor does it prevent, a two-vehicle crash.

Two-vehicle crashes involving the rear end of one or more vehicles and sideswipe same-direction crashes have been eliminated from the study. Two-vehicle rear-end and sideswipe same-direction crashes might be meaningful choices for comparison crashes because they share similar vision-related causal factors as the target crashes, even though DRLs could play a role as a countermeasure in rear end crashes. One problem is that the number of such crashes is much smaller than single vehicle crashes and the results would not have enough power to reject the null hypothesis. However, there is another argument that although rear-end and same-direction sideswipe crashes are not the intended target of DRLs, they are relevant since they draw attention to following vehicles – particularly tailgating vehicles – where drivers may respond with actions that potentially can increase or decrease the risk of a crash. If this is the case, design issues of location, brightness and color may be relevant.

Crashes of three or more vehicles were eliminated from the analysis. The crash geometry can become quite complex and vague for crashes of three or more vehicles and the number of such crashes is small. It is easy to misclassify such a crash. Therefore, to reduce the possibility of contamination of the analysis, all crashes involving three or more vehicles have been eliminated.

Another possible source of contamination, albeit a small one, is crashes involving parked vehicles in a fatal crash. To insure a vehicle involved in the crash was not parked, the requirement that a driver was present or that the driver had left the scene, was imposed.

The vehicles in the analysis were restricted to passenger vehicles of model year 1995 and later. Passenger vehicles include passenger cars, SUVs, light tucks, and vans. The target group of vehicles with daytime running lamps and the comparison group of vehicles without daytime running lamps were identified by analysis of the Vehicle Identification Number, VIN. Analysis of the VIN partitioned vehicles into 4 distinct groups: 1) vehicles that had DRLs as standard equipment, 2) vehicles that did not have DRLs as standard equipment nor as a standard option,
3) vehicles that have DRLs as a standard option, and 4) other vehicles including vehicles where the VIN was not reported or could not be decoded.

The target group of vehicles was the group of vehicles with DRLs as standard equipment. The comparison group of vehicles was the group of vehicles without DRLs, which did not have DRLs as a standard option. Vehicles with DRLs as a standard option and the vehicles in the "other" category were eliminated from the analysis.

Caveats

To analyze the effect of a new vehicle safety device one needs to compare it to vehicles that do not have the device and in situations that should and should not be affected by the device. One attempts to assure that the respective partition of vehicles and crashes eliminates any lurking variables, but this can never be fully guaranteed. The selection and partition of vehicles and crashes were based on the analytic judgment.

DRL Effectiveness in Fatal Two-Vehicle Crashes

The target crashes are two-vehicle crashes where the vehicles are traveling in opposite-directions. The target crashes include head-on, and sideswipe opposite direction crashes. Single-vehicle crashes are the comparison crashes.

Table 2 shows the cross tabulation of the target and single-vehicle crashes under daytime and nighttime conditions for vehicles equipped with DRLs.

Table 2
DRL-Equipped Vehicles in Target and
Single-Vehicle Fatal Crashes, FARS 1995-2001

Time of Day	Target Crashes	Single-Vehicle Crashes	Total
Daytime	2,117	3,360	5,477
Nighttime	1,047	4,573	5,620
Total	3,164	7,933	11,097
Source: NHTSA, NCSA, FARS			

Table 3 shows the cross tabulation of the target and single-vehicle crashes under daytime and nighttime conditions for the comparison group of vehicles without DRLs.

NCSA

Table 3
Vehicles w/o DRL in Target and
Single-Vehicle Fatal Crashes, FARS 1995-2001

Time of Day	Target Crashes	Single-Vehicle Crashes	Total
Daytime	6,699	10,058	16,757
Nighttime	3,450	13,413	16,863
Total	10,149	23,471	33,620
Source: NHTSA, NCSA, FARS			

DRL Effectiveness in Fatal Two-Vehicle Crashes - Results

The effectiveness, based on the simple odds, of DRLs in preventing two-vehicle opposite direction fatal crashes during daylight is estimated to be **5.3** percent with ($p = 0.052$).

Passenger vehicle type may influence the effectiveness of DRLs. To examine this issue, vehicle types were included in the logistic fit of the data. The results are similar. The effectiveness, based on the simple odds, of DRLs in preventing two-vehicle opposite direction crashes during daylight is estimated to be **5.1** percent with ($p = 0.061$) when adjusting for vehicle type.

DRL Effectiveness in Non-Fatal Two-Vehicle Crashes

The target crashes are two-vehicle crashes where the vehicles are traveling in opposite-directions. Single-vehicle crashes are the comparison crashes.

Table 4 shows the cross tabulation of the target and single-vehicle non-fatal crashes under daytime and nighttime conditions for vehicles equipped with DRLs. Since NASS/GES is a complex sample survey a program such as SUDAAN must be used to estimate the level of significance of the parameters.

NCSA

Table 4
DRL-Equipped Vehicles in Target and
Single -Vehicle Non-Fatal Crashes, NASS/GES 1995-2001

Time of Day	Target Crashes	Single-Vehicle Crashes	Total
Daytime	972,000	248,000	1,220,000
Nighttime	215,000	216,000	432,000
Total[25]	1,188,000	464,000	1,652,000
Source: NHTSA, NCSA, NASS/GES			

Table 5 shows the cross tabulation of the target and single-vehicle non-fatal crashes under daytime and nighttime conditions for the comparison group of vehicles without DRLs.

Table 5
Vehicles w/o DRL in Target and
Single-Vehicle Non-Fatal Crashes, NASS/GES 1995-2001

Time of Day	Target Crashes	Single-Vehicle Crashes	Total
Daytime	3,074,000	737,000	3,812,000
Nighttime	695,000	608,000	1,303,000
Total[24]	3,770,000	1,345,000	5,115,000
Source: NHTSA, NCSA, NASS/GES			

DRL Effectiveness in Non-Fatal Two-Vehicle Crashes - Results

The effectiveness, based on the simple odds, of DRLs in preventing two-vehicle opposite direction non-fatal crashes during daylight is estimated to be **5.2** percent with ($p = 0.075$).

Passenger vehicle type may influence the effectiveness of DRLs. To examine this issue, vehicle types were included in the logistic fit of the data. The results are similar. The effectiveness of DRLs in preventing two-vehicle opposite direction non-fatal crashes during daylight is estimated to be **4.4** percent with ($p = 0.133$) when adjusting for vehicle type. Since the value of p is greater than 0.1, the null hypothesis cannot be rejected. However, since this estimate of effectiveness is similar to the significantly significant value calculated without adjusting for vehicle type, with ($p = 0.075$) one could interpret this estimate as a weak confirmation of the previous result.

---

[25] Totals may not add due to rounding.

DRL Effectiveness in Fatal Single-Vehicle Pedestrian/Cyclist Crashes

As daytime running lamps reduce two-vehicle opposite-direction crashes, daytime running lamps may also reduce single-vehicle crashes with pedestrians or cyclists. To answer that question, one can modify the approach used above. FARS, 1995 to 2001, can again be used for this analysis. However, the analysis is performed at the person level, rather than the vehicle level[26]. The target group of persons is fatally injured pedestrians and cyclists in single vehicle crashes; the comparison group of persons is fatally injured occupants in single vehicle crashes. The target time period is daytime, including dawn and dusk and the comparison time period is night. The results follow:

Table 6
Single-vehicle Pedestrian and Cyclist Fatalities FARS 1995-2001
Vehicles Equipped with DRLs

Time of Day	Pedestrian and Cyclist Deaths	Occupant Deaths	Total
Daytime	710	6,288	6,998
Nighttime	1,153	8,136	9,289
Total	1,863	14,424	16,287
Source: NHTSA, NCSA, FARS			

Table 7
Single-vehicle Pedestrian and Cyclist Fatalities FARS 1995-2001
Vehicles Not Equipped with DRLs

Time of Day	Pedestrian and Cyclist Deaths	Occupant Deaths	Total
Daytime	2,515	19,540	22,055
Nighttime	3,876	24,946	28,822
Total	6,391	44,486	50,877
Source: NHTSA, NCSA, FARS			

---

[26] It is possible for a pedestrian fatality and an occupant fatality to occur in the same crash. In this case, both the pedestrian and cyclist death cell and occupant death cell are incremented. To avoid potential single vehicle crashes involving a pedestrian/cyclist death and an occupant death from confounding the data, this analysis is performed at the person level, not the crash level.

DRL Effectiveness in Fatal Single-Vehicle Pedestrian/Cyclist Crashes - Results

The effectiveness, based on the simple odds, of DRLs in preventing single-vehicle pedestrian/cyclist fatalities during daylight is estimated to be **12.4** percent with ($p = 0.002$).

Passenger vehicle type may influence the effectiveness of DRLs. To examine this issue, vehicle types were included in the logistic fit of the data. The results are similar. The effectiveness, based on the simple odds, of DRLs in preventing single-vehicle pedestrian/cyclist fatal crashes during daylight is estimated to be **12.9** percent with ($p = 0.002$) when adjusting for vehicle type.

DRL Effectiveness in Fatal Crashes of a Passenger Vehicle with a Motorcycle

Target crashes are two-vehicle opposite direction crashes between a passenger vehicle and a motorcycle. Comparison crashes are single vehicle crashes. In the analysis that follows, the DRL status of the passenger vehicle involved in a two-vehicle crash with a motorcycle determined if the crash was a DRL equipped crash or a non-DRL equipped crash.

Table 8 shows the cross tabulation of the target and single passenger vehicle crashes under daytime and nighttime conditions for passenger vehicles equipped with DRLs.

Table 8
Passenger Vehicles with DRLs Involved in Fatal 2-Vehicle Crashes of a Motorcycle and a Single Passenger Vehicle, FARS 1995-2001

Time of Day	Target 2-Vehicle Motorcycle Crashes	Single Passenger Vehicle Crashes	Total
Daytime	62	3,360	3,422
Nighttime	30	4,573	4,603
Total	92	7,933	8,025
Source: NHTSA, NCSA, FARS			

Table 9 shows the cross tabulation of the target and single passenger vehicle crashes under daytime and nighttime conditions for the comparison group of passenger vehicles without DRLs.

NCSA

Table 9
Passenger Vehicles w/o DRLs Involved in Fatal 2-Vehicle Crashes of a Motorcycle and a Single Passenger Vehicle, FARS 1995-2001

Time of Day	Target 2-Vehicle Motorcycle Crashes	Single Passenger Vehicle Crashes	Total
Daytime	239	10,058	10,297
Nighttime	86	13,413	13,499
Total	325	23,471	23,796
Source: NHTSA, NCSA, FARS			

DRL Effectiveness in Fatal Crashes of a Passenger Vehicle with a Motorcycle - Results

The effectiveness, based on the simple odds, of DRLs in preventing two-vehicle opposite direction crashes between a passenger vehicle and a motorcycle during daylight is estimated to be **23.2** percent with ($p = 0.065$).

Passenger vehicle type may influence the effectiveness of DRLs. To examine this issue, vehicle types were included in the logistic fit of the data. The results are similar. The effectiveness, based on the simple odds, of DRLs in preventing two-vehicle opposite direction crashes between a passenger vehicle and a motorcycle during daylight is estimated to be **22.6** percent with ($p = 0.074$) when adjusting for vehicle type.

Conclusions

The effectiveness of daytime running lamps, based on the simple odds, was analyzed in the preceding sections using data from FARS and NASS/GES from calendar years 1995 to 2001. FARS and NASS/GES data show that during the period of the study 1995 to 2001, DRLs reduced daylight two passenger vehicle opposite-direction crashes by about 5 percent. DRLs have also been shown to reduce fatal opposite direction crashes between a motorcycle and a passenger vehicle by 23 percent. The results for two-vehicle daytime opposite-direction crashes are statistically significant at the $p < 0.10$ level, although one would prefer a statistical level of $p < 0.05$.

FARS data were also used to estimate the effectiveness, based on the simple odds, of DRLs in reducing pedestrian/cyclist fatalities in single-vehicle fatal crashes. The analysis shows that DRLs reduced pedestrian/cyclist fatalities by more than 12 percent. These results are highly significant at a statistical level of $p = 0.002$.

This analysis is based on US historical data and does not reflect what will happen in the future. The techniques used do not predict the crash reducing effectiveness of DRLs if the entire fleet is equipped with DRLs nor if drivers become habituated to DRLs. These are limitations of historical crash data.

As additional data become available it may be appropriate to further investigate the effectiveness of DRLs in a variety of crash configurations including pedestrian and motorcycle crashes.

Appendix A

Analytic Approach

The primary analytic approach used to estimate the effectiveness, E, of daytime running lamps is based on the generalized simple odds. The effectiveness, based on the simple odds approach, is defined as:

$$E = 1 - e^{\beta}$$

Where ß is the coefficient of the following equation:

$$TC\_DT = \beta * DRL + \sum_i \theta_i * X_i + error$$

Where: TC_DT = 1 if the crash is a target crash that occurred during the day, 0 otherwise and DRL = 0 if the vehicle has DRLs, otherwise 1. A bivariate logistic fit of the data is calculated using a maximum likelihood estimate. FARS data can be analyzed using SAS®, however, since NASS/GES data come from a complex survey rather than a census, SUDAAN had to be used to estimate the variance and significance of the estimated coefficients.

In the event that one does not need to control for variables such as vehicle type, the X_i terms are zero and an arithmetic approach to calculate the effectiveness exists. In this case, the effectiveness, E is equivalent to:

$$E = 1 - (O_{DRL}/O_{CMP})$$

Where

$$O = TD/(CD+TN+CN)$$

and is evaluated for both the vehicles equipped with DRLs, O_{DRL}, and the vehicles in the comparison group without DRLs, O_{CMP}.

TD   is the number of vehicles/persons in Targeted crashes during Daylight.

CD   is the number of vehicles/persons in Comparison crashes during Daylight.

TN   is the number of vehicles/persons in Targeted crashes at Night.

CN   is the number of vehicles/persons in Comparison crashes at Night.

In this simplified case, for FARS data, the variance of ln (1-E), can be estimated as the sum of the squared of the reciprocals of the four groups of observations. That is:

$$VAR\ [\ln(1-E)] \sim [1/TD_{DRL}]^2 + [1/(CD_{DRL} + TN_{DRL} + CN_{DRL})]^2 +$$
$$[1/TD_{CMP}]^2 + [1/(CD_{CMP} + TN_{CMP} + CN_{CMP})]^2$$

This technique to estimate the variance of the $\ln(1-E)$ does not apply to weighted survey data, which requires complex software such as SUDAAN.

Logistic Regression Estimates Using the Simple Odds

Note that, with the exception of Table A-4, the value of p for the coefficient of DRL is <0.1.

Table A-1					
DRL Effectiveness in Fatal Two-Vehicle Crashes Based on Simple Odds					
Parameter	Odds Ratio	Estimate	Standard Error	Wald Chi-Square	Pr > ChiSq p
Intercept	N/A	1.4450	0.0242	3,577.07	<0.0001
DRL	0.947	-0.0541	0.0278	3.79	0.0515
Source: NHTSA, NCSA, FARS, SAS®					

Table A-2					
DRL Effectiveness in Fatal Two-Vehicle Crashes / Adjusted for Vehicle Type Based on Simple Odds					
Parameter	Odds Ratio	Estimate	Standard Error	Wald Chi-Square	Pr > ChiSq p
Intercept	N/A	1.4902	0.0270	3,053.61	< 0.0001
DRL	0.949	-0.0523	0.0279	3.52	0.0606
Sport Utility	1.211	0.1917	0.0356	29.01	<0.0001
Van	0.475	-0.2938	0.0442	44.11	<0.0001
Light Trucks	0.817	-0.2025	0.0283	51.10	<0.0001
Source: NHTSA, NCSA, FARS, SAS®					

NCSA

Table A-3
DRL Effectiveness in Non-Fatal Two-Vehicle Crashes Based on Simple Odds

Parameter	Odds Ratio	Estimate	Standard Error	Wald Chi-Square	Pr > ChiSq p
Intercept	N/A	-0.3579	0.05	3.33	0.0751
DRL	0.948	-0.0529	0.03	3.33	0.0751
Source: NHTSA, NCSA, NASS/GES, SUDAAN					

Table A-4
DRL Effectiveness in Non-Fatal Two-Vehicle Crashes / Adjusted for Vehicle Type Based on Simple Odds

Parameter	Odds Ratio	Estimate	Standard Error	Wald Chi-Square	Pr > ChiSq p
Intercept	N/A	-0.3779	0.13	3.74	0.0017
DRL	0.956	-0.0445	0.03	2.34	0.1333
Sport Utility	1.099	0.0941	0.05	4.30	0.0441
Van	0.826	-0.1906	0.08	6.13	0.0173
Light Trucks	1.089	0.0856	0.05	3.53	0.0672
Source: NHTSA, NCSA, NASS/GES, SUDAAN					

Table A-5
DRL Effectiveness in Fatal Single-Vehicle – Pedestrian/Cyclist Crashes Based on Simple Odds

Parameter	Odds Ratio	Estimate	Standard Error	Wald Chi-Square	Pr > ChiSq p
Intercept	N/A	3.0883	0.0384	6,476.45	<0.0001
DRL	0.876	-0.1318	0.0435	9.19	0.0024
Source: NHTSA, NCSA, FARS, SAS®					

Table A-6
DRL Effectiveness in Fatal Single-Vehicle – Pedestrian/Cyclist Crashes Adjusted for Vehicle Type Based on Simple Odds

Parameter	Odds Ratio	Estimate	Standard Error	Wald Chi-Square	Pr > ChiSq p
Intercept	N/A	3.1449	0.0427	5,424.56	<0.0001
DRL	0.871	-0.1377	0.0437	9.94	0.0016
Sport Utility	1.231	0.2082	0.0527	15.58	<0.0001
Van	0.812	-0.2086	0.0586	12.69	0.0004

| Light Trucks | 0.752 | -0.2853 | 0.0445 | 41.08 | <0.0001 |

Source: NHTSA, NCSA, FARS, SAS®

Table A-7
DRL Effectiveness for Two-Vehicle Fatal Crashes
Involving a Motorcycle and a Passenger Vehicle Based on Simple Odds

Parameter	Odds Ratio	Estimate	Standard Error	Wald Chi-Square	Pr > ChiSq p
Intercept	N/A	4.8552	0.1275	1,450.52	<0.0001
DRL	0.768	-0.2645	0.1431	3.42	0.0645

Source: NHTSA, NCSA, FARS, SAS®

Table A-8
DRL Effectiveness for Two-Vehicle Fatal Crashes
Involving a Motorcycle and a Passenger Vehicle Adjusted for Vehicle Type
Based on Simple Odds

Parameter	Odds Ratio	Estimate	Standard Error	Wald Chi-Square	Pr > ChiSq p
Intercept	N/A	5.1603	0.1478	1,224.38	<0.0001
DRL	.0774	-0.2564	0.1436	3.19	0.0741
Sport Utility	0.627	-0.4664	0.1572	8.80	0.0030
Van	0.499	-0.6960	0.2044	11.60	0.0007
Light Trucks	0.557	-0.5846	0.1408	17.25	<0.0001

Source: NHTSA, NCSA, FARS, SAS®

NCSA

Appendix B

Alternate Analytic Approach

This section is included at the request of the reviewers of the paper. The odds ratio is easier to understand for inexperienced analysts than the simple odds and, like the simple odds, attempts to control for a variety of factors other than the presence or absence of DRLs. Unfortunately, when using the odds ratio, the estimated effectiveness of DRLs is extremely sensitive to small changes encountered in real world crash data and none of the results were statistically significant. This does not mean that DRLs do not reduce target crashes during the daytime. It just means that the odds ratio technique does not detect these changes over the inherent background noise of the data system.

The effectiveness, based on the odds ratio, is defined as:

$$E = 1 - e^{\beta}$$

Where ß is the coefficient of the following equation:

$$LIGHT = \beta*(DRL \times CRASH) + a_1*DRL + a_2*CRASH + S_i\ ?_i*X_i + error$$

Where: LIGHT = 1 if the crash occurred during the day, 0 otherwise
          DRL = 0 if the vehicle has DRLs, otherwise 1.
          CRASH = 1 if the crash is a target crash, and 0 if the crash is a comparison crash.

A bivariate logistic fit of the data is calculated using a maximum likelihood estimate. FARS data can be analyzed using SAS®, however, since NASS/GES data come from a complex survey rather than a census, SUDAAN had to be used to estimate the variance and significance of the estimated coefficients.

In the event that one does not need to control for variables such as vehicle type, the X_i terms are zero and an arithmetic approach to calculate the effectiveness exists. In this case, the effectiveness, E is equivalent to:

$$E = 1 - (?_{DRL}/?_{CMP})$$

Where

$$? = (TD/CD)/(TN/CN)$$

and is evaluated for both the vehicles equipped with DRLs, $?_{DRL,}$ and the vehicles in the comparison group without DRLs, $?_{CMP}$.

In this simplified case, for FARS data, the variance of ln (1-E), can be estimated as the sum of the squares of the reciprocals of the eight groups of observations. That is:

 National Center for Statistics and Analysis ♦ 400 Seventh St., S.W., Washington, D.C. 20590   23

$$VAR\ [\ln(1-E)]\ \tilde{}\ [1/TD_{DRL}]^2 + [1/CD_{DRL}]^2 + [1/TN_{DRL}]^2 + [1/CN_{DRL}]^2 +$$
$$[1/TD_{CMP}]^2 + [1/CD_{CMP}]^2 + [1/TN_{CMP}]^2 + [1/CN_{CMP}]^2$$

Note that VAR [$\ln(1-E)$] is much larger for the odds ratio than for the simple odds. As a result, the values of p, for each of the evaluated crash types in this study, are larger than 0.1, therefore the null hypothesis cannot be rejected and there is no reason to believe that the results, based on the odds ratio, did not occur by chance.

Using the data of Tables 2 and 3, the estimates of effectiveness of DRLs are calculated using the odds ratio. The result for two-vehicle opposite direction fatal crashes is –6.3 percent with (p=**0.229**). When adjusting for vehicle type, the result is –6.3 percent with (p=**0.235**). The values of p in both cases are larger than 0.1, therefore the null hypothesis cannot be rejected and there is no reason to believe that the results, based on the odds ratio, did not occur by chance.

Using the data of Tables 4 and 5, the estimates of effectiveness of DRLs are calculated using the odds ratio. The result for two-vehicle opposite direction non-fatal crashes is –7.9 percent with (p=**0.186**). When adjusting for vehicle type, the result is –7.6 percent with (p=**0.202**). The values of p in both cases are larger than 0.1, therefore the null hypothesis cannot be rejected and there is no reason to believe that the results, based on the odds ratio, did not occur by chance.

Using the data of Tables 6 and 7, the estimates of effectiveness of DRLs are calculated using the odds ratio. The result for fatal single-vehicle pedestrian/cyclist crashes is 3.8 percent with (p=**0.498**). When adjusting for vehicle type, the result is 4.6 percent with (p=**0.415**). The values of p in both cases are larger than 0.1, therefore the null hypothesis cannot be rejected and there is no reason to believe that these results, based on the odds ratio, did not occur by chance.

Using the data of Tables 8 and 9, the estimates of effectiveness of DRLs are calculated using the odds ratio. The result for crashes of a passenger vehicle with a motorcycle is 26.0 percent with (p=**0.284**). When adjusting for vehicle type, the result is 22.0 percent with (p=**0.335**). The values of p in both cases are larger than 0.1, therefore the null hypothesis cannot be rejected and there is no reason to believe that the results, based on the odds ratio, did not occur by chance.

NCSA

Logistic Regression Estimates Using the Odds Ratio

Note that the value of p for the coefficient of DRLxCRASH is always larger than 0.1. Therefore the null hypothesis cannot be rejected and the estimates, based on the odds ratio, do not improve our understanding of the effectiveness of DRLs. However, if the estimate of effectiveness is larger than 20 percent, the estimates, based on the odds ratio, are similar to the estimates calculated using the simple odds, albeit not statistically significant.

<table>
<tr><th colspan="6">Table B-1
DRL Effectiveness in Fatal Two-Vehicle Crashes Based on Odds Ratio</th></tr>
<tr><th>Parameter</th><th>Odds Ratio</th><th>Estimate</th><th>Standard Error</th><th>Wald Chi-Square</th><th>Pr > ChiSq
p</th></tr>
<tr><td>Intercept</td><td>N/A</td><td>0.3082</td><td>0.227</td><td>184.01</td><td><0.0001</td></tr>
<tr><td>DRLxCRASH</td><td>1.063</td><td>0.0608</td><td>0.0506</td><td>1.45</td><td>0.2291</td></tr>
<tr><td>DRL</td><td>0.980</td><td>-0.0204</td><td>0.0263</td><td>0.60</td><td>0.4381</td></tr>
<tr><td>CRASH</td><td>0.363</td><td>0.3082</td><td>0.0441</td><td>527.08</td><td><0.0001</td></tr>
<tr><td colspan="6">Source: NHTSA, NCSA, FARS, SAS®</td></tr>
</table>

<table>
<tr><th colspan="6">Table B-2
DRL Effectiveness in Fatal Two-Vehicle Crashes / Adjusted for Vehicle Type
Based on Odds Ratio</th></tr>
<tr><th>Parameter</th><th>Odds Ratio</th><th>Estimate</th><th>Standard Error</th><th>Wald Chi-Square</th><th>Pr > ChiSq
p</th></tr>
<tr><td>Intercept</td><td>N/A</td><td>0.4052</td><td>0.0247</td><td>269.55</td><td>< 0.0001</td></tr>
<tr><td>DRLxCRASH</td><td>1.063</td><td>0.0606</td><td>0.0507</td><td>1.43</td><td>0.2315</td></tr>
<tr><td>DRL</td><td>0.998</td><td>-0.0019</td><td>0.0264</td><td>0.01</td><td>0.9435</td></tr>
<tr><td>CRASH</td><td>0.361</td><td>-1.0193</td><td>0.0442</td><td>531.82</td><td><0.0001</td></tr>
<tr><td>Sport Utility</td><td>0.738</td><td>-0.3036</td><td>0.0273</td><td>123.75</td><td><0.0001</td></tr>
<tr><td>Van</td><td>0.604</td><td>-0.5044</td><td>0.0689</td><td>167.83</td><td><0.0001</td></tr>
<tr><td>Light Trucks</td><td>0.920</td><td>-0.0836</td><td>0.0237</td><td>12.43</td><td><0.0004</td></tr>
<tr><td colspan="6">Source: NHTSA, NCSA, FARS, SAS®</td></tr>
</table>

NCSA

Table B-3
DRL Effectiveness in Non-Fatal Two-Vehicle Crashes Based on Odds Ratio

Parameter	Odds Ratio	Estimate	Standard Error	Wald Chi-Square	Pr > ChiSq p
Intercept	N/A	-0.1343	0.08	222.88	<0.0001
DRLxCRASH	1.079	0.0763	0.06	1.81	0.1259
DRL	0.944	-0.0574	0.05	1.56	0.2184
CRASH	0.253	-1.3725	0.07	383.31	<0.0001

Source: NHTSA, NCSA, NASS/GES, SUDAAN

Table B-4
DRL Effectiveness in Non-Fatal Two-Vehicle Crashes / Adjusted for Vehicle Type Based on Odds Ratio

Parameter	Odds Ratio	Estimate	Standard Error	Wald Chi-Square	Pr > ChiSq p
Intercept	N/A	-0.0957	0.08	141.85	<0.0000
DRLxCRASH	1.076	0.0735	0.06	1.68	0.2024
DRL	-0.954	-0.0467	0.05	0.97	0.3294
CRASH	0.253	-1.3728	0.07	387.03	<0.0000
Sport Utility	0.921	-0.0825	0.04	5.20	0.0276
Van	0.718	-0.3320	0.06	33.31	<0.0000
Light Trucks	1.089	0.0856	0.03	5.05	0.0237

Source: NHTSA, NCSA, NASS/GES, SUDAAN

NCSA

Table B-5
DRL Effectiveness in Fatal Single-Vehicle – Pedestrian/Cyclist Crashes
Based on Odds Ratio

Parameter	Odds Ratio	Estimate	Standard Error	Wald Chi-Square	Pr > ChiSq p
Intercept	N/A	0.2577	0.0168	235.46	<0.0001
DRLxPERSON	0.962	-0.0389	0.0575	0.46	0.4984
DRL	0.987	-0.0134	0.0193	0.42	0.4877
PERSON	0.876	-0.1318	0.0435	9.19	0.0024

Source: NHTSA, NCSA, FARS, SAS®

Table B-6
DRL Effectiveness in Fatal Single-Vehicle – Pedestrian/Cyclist Crashes
Adjusted for Vehicle Type Based on Odds Ratio

Parameter	Odds Ratio	Estimate	Standard Error	Wald Chi-Square	Pr > ChiSq p
Intercept	N/A	0.4387	0.0186	554.10	<0.0001
DRLxPERSON	0.954	-0.0472	0.0579	0.66	0.4153
DRL	1.031	0.0309	0.0195	2.51	0.1133
PERSON	1.238	0.2137	0.0509	17.62	<0.0001
Sport Utility	0.619	-0.4790	0.0208	532.62	<0.0001
Van	0.461	-0.7749	0.0269	831.38	<0.0001
Light Trucks	0.846	-0.1667	0.0206	65.46	<0.0001

Source: NHTSA, NCSA, FARS, SAS®

Table B-7
DRL Effectiveness for Two-Vehicle Fatal Crashes
Involving a Motorcycle and a Passenger Vehicle Based on Odds Ratio

Parameter	Odds Ratio	Estimate	Standard Error	Wald Chi-Square	Pr > ChiSq p
Intercept	N/A	0.3082	0.0227	184.01	<0.0001
DRLxCRASH	0.760	-0.2851	0.2568	1.15	0.2842
DRL	0.980	-0.0204	0.0263	0.60	0.4381
CRASH	0.356	-1.0341	0.2236	21.40	<0.0001

Source: NHTSA, NCSA, FARS, SAS®

Table B-8
DRL Effectiveness for Two-Vehicle Fatal Crashes
Involving a Motorcycle and a Passenger Vehicle Adjusted for Vehicle Type
Based on Odds Ratio

Parameter	Odds Ratio	Estimate	Standard Error	Wald Chi-Square	Pr > ChiSq p
Intercept	N/A	0.4268	0.0254	282.23	<0.0001
DRLxCRASH	0.780	-0.2485	0.2577	0.93	0.3349
DRL	1.003	0.0030	0.0264	0.01	0.9087
CRASH	0.355	-1.0363	0.2242	21.36	<0.0001
Sport Utility	0.691	-0.3695	0.0309	143.26	<0.0001
Van	0.529	-0.6362	0.0460	190.93	<0.0001
Light Trucks	0.905	-0.0993	0.0283	12.27	0.0005

Source: NHTSA, NCSA, FARS, SAS®

Appendix C

The following SAS® code was used to partition FARS 1996 vehicle crashes. The code for the NASS/GES is similar.

```
/* COMPARISON CRASHES SINGLE VEHICLE CRASHES */

LIBNAME FARS96 'L:\FARSSAS\FARS96';

DATA CRASH;
  SET FARS96.ACCIDENT(KEEP = ST_CASE LGT_COND VE_FORMS MAN_COLL
WEATHER);

LENGTH TGT_CRSH $8;

*  IF TWO VEHICELES CRASH AND;
*  HEAD-ON OR SIDESWIPE DIFFERENT DIRECTIONS;

IF (VE_FORMS EQ 2) AND
        ((2 EQ MAN_COLL) OR (6 EQ MAN_COLL))
        THEN TGT_CRSH ='MUL TGT';

/* ELSE SINGLE VEHICLE CRASHES */
ELSE IF (VE_FORMS EQ 1) THEN TGT_CRSH = 'SINGLE';
ELSE DELETE;

*DEFINE THE DICHOTOMOUS VARIABLE D_CRASH;

IF (VE_FORMS EQ 2) AND
        ((2 EQ MAN_COLL) OR (6 EQ MAN_COLL))
        THEN D_CRASH = 1;

/* ELSE SINGLE VEHICLE CRASHES */
ELSE IF (VE_FORMS EQ 1) THEN D_CRASH = 0;
ELSE DELETE;

LENGTH LIGHT $7;

*IF DAYLIGHT DAWN OR DUSK;
IF (LGT_COND EQ 1 OR 4 LE LGT_COND LE 5) THEN LIGHT = 'DAYTIME';

*IF DARK OR DARK AND LIGHTED;
ELSE IF (2 LE LGT_COND LE 3) THEN LIGHT = 'NIGHT';
```

```
ELSE DELETE;

*  DEFINE THE DICHOTOMOUS VARIABLE D_LIGHT;
IF (LGT_COND EQ 1 OR 4 LE LGT_COND LE 5) THEN D_LIGHT = 1;
ELSE IF (2 LE LGT_COND LE 3) THEN D_LIGHT = 0;

* DEFINE THE DICHOTOMOUS VARIABLE MUL_DAY;
* THIS IS FOR THE SIMPLE ODDS CALCULATION;

IF (D_CRASH = 1 AND D_LIGHT = 1) THEN MUL_DAY =1;
ELSE MUL_DAY = 0;
```

Bibliography

Allen, M.J. and Clark, J.R. Automobile Running Lights - A Research Report. American Journal of Optometry and Archives of American Academy of Optometry 41(5).293-315, 1964.

Andersson, K.; Nilsson, G.; and Salusjärvi, S. The Effect On Traffic Accidents on the Recommended use of Vehicle Running Lights in the Daytime in Finland. Report No 102. Swedish Road and Traffic Research Institute (VTI), 1976.

Andersson, K.; Nilsson, G. The Effects on Accidents of Compulsory use of Running Lights During Daylight in Sweden. Report No. 208A, Swedish Road and Traffic Research Institute (VRI),1981.

Arora, H.; Collard, D.; Robbins, G.; Welbourne, E.R.; White, J.G. Effectiveness of Daytime Running Lights in Canada, Report No. TP12298 (E). Ottawa, Canada, Transport Canada 1994.

Attwood, D.A. Daytime Running Lights Project I: Research Program and Preliminary Results. Report No. RSU 75/1, Transport Canada, 1975.

Attwood, D.A. Daytime Running Lights Project II: Vehicle Detection as a Function of Headlight use and Ambient Illumination. Report No. RSU 75/2, Transport Canada, 1975.

Attwood, D.A. and Angus, R.G. Daytime Running Lights Project III: Pilot Validation Study of Field Detection Experiments. Report No. RSU 75/3, Transport Canada, 1975.

Attwood, D.A. Daytime Running Lights Project IV: Two-lane Passing Performance as a Function of Headlight Intensity and Ambient Illumination. Report No. RSU 76/1, Transport Canada, 1976.

Attwood, D.A. The Potential of Daytime Running Lights as a Vehicle Collision Countermeasure. SAE Technical Paper 810190. Society of Automotive Engineers, 1981.

Burger, W.; Smith R.; Ziedman, K. Evaluation of the Conspicuity of Daytime Running Lights. DOT HS 807 609, April 1990.

Cantilli, E.J. Daylight "Running Lights" Reduce Accidents. Traffic Engineering, 39(5), 52-57, 1969.

Claton, A.B. and Mackay, G.M. Aetiology of Traffic Accidents. Health Bulletin, 31(4), 277-280, 1972.

Dahlstedt, S. and Rumar, K. Vehicle Colour and Front Conspicuity in Some Simulated Rural Traffic Situations. University of Uppsala, Department of Psychology, Sweden, 1973.

Elvik, R. The Effects of Accidents of Compulsory Use of Daytime Running Lights for Cars in Norway. Accident Analysis and Prevention, 25(4) 383-398, 1993.

Farmer, C.M. and Williams, A.F. Effects of daytime running lights on multiple-vehicle daylight crashes in the United States; Accident Analysis. and Prevention, Volume 34, pp 197-203, 2002.

Fleiss, J. L. Statistical Methods for Rates and Proportions, Second edition. New York, John Wiley and Sons, 1981.

Hansen, L.K. Daytime Running Lights in Denmark - Evaluation of the Safety Effect. Translated exact from Technical Report 2/1993. Copenhagen: Danish Council of Road Safety Research, 1993.

Hauer, E. Before-and-After Studies in Road Safety, Estimating the Effect of Highway and Traffic Engineering Measures on Road Safety. Lecture notes, March 1995. Department of Civil Engineering, University of Toronto, Ontario, Canada, 1995.

Highway Loss Data Institute Bulletin Volume 15, Number 1, December 1997.

Hills, B.L. Vision, Visibility and Perception in Driving. Perception, 9, 183-216, 1980.

Hollo, P. Changes of the DRL-Regulations and their Effects on Traffic Safety in Hungary. Paper presented at the conference: Strategic Highway Safety Program and Traffic Safety, the Czech Republic, September 20-22, 1995. Preprint for sessions on September 21, 1995.

Hollo, P., Changes in the Legislation on the use of Daytime Running Lights by Motor Vehicles and Their Effect on Road Safety in Hungary, Accid. Anal. And Prev., Vol 30, No.2, pp 183-199, 1998.

Koornstra, M. J., Bijleveld, F. and Hagenzieker, M., The Safety Effects of Daytime Running Lights, SWOV Institute for Road Safety Research, Leidschendam The Netherlands, 1997.

Kirkpatrick, M. and Marshall, R. Evaluation of Glare From Daytime Running Lights. DOT HS 807 609, October 1989.

Lau, E., Daytime Running Light Effectiveness A Preliminary Evaluation, Exponent Inc. Menlo Park, CA , presented at a Government/Industry Meeting, Washington DC, June 19-21, 2000.

Sekuler, R. and Blake, R. Perception, (Second Edition) Toronto: McGraw-Hill, 1990.

Stein, H. Fleet Experience with Daytime Running Lights in the United States. SAE Technical Paper 851239. Warrendale, PA, Society of Automotive Engineers, 1985.

Tessmer, J. M. FARS Analytic Reference Guide 1975-2002, DOT HS 808 463, U.S. Department of Transportation, Washington D.C., June 2002.

Tessmer, J.M. A Preliminary Assessment of the Crash-Reducing Effectiveness of Passenger Car Daytime Running Lamps (DRLs), DOT HS 808 645 June 2000.

Theeuwes, J. and Riemersma, J. Daytime Running Lights as a Vehicle Collision Countermeasure: The Swedish Evidence Reconsidered. Accident Analysis and Prevention. 27:633-642, 1995.

Tofflemire, T. C., Whitehead, P.C. An Evaluation of the Impact of Daytime Running Lights on Traffic Safety in Canada, Journal of Safety Research, Vol 28, No. 4, 1997.

Thompson, P.A. Daytime Running Lamps (DRLs) for Pedestrian Protection SAE Paper 2003-0102072, April 2003.

Williams, A. F. and Lancaster, K. A. The Prospects of Daytime Running Lights for Reducing Vehicle Crashes in the United States. Public Health Reports, Vol. 110, No. 3 233-239, 1995.